For Ed Wilson
who helps us
"stay alive in the
vanished forests of
the world"
With gratitude

Jonathan
Kingdon

KILIMANJARO: ANIMALS IN A LANDSCAPE

JONATHAN KINGDON, born at Tabora in Tanganyika, had his primary schooling in Tanganyika and his
secondary education at Oxford. His upbringing was domi-
nated by three influences; the African wilderness, natural
history and drawing.

After an education in science and art, he returned to
East Africa to lecture in Fine Art at Makerere University.
There he began working on *East African Mammals*, a
seven-volume atlas of evolution in Africa that won him
international renown. His versatility, however, does not
allow him to be tied to one place or to a single path of
knowledge. He is a consultant ecologist and has turned his
mind to palaeontology, archaeology and anthropology.

His preoccupation with colour and form led him to
investigate the origin of facial patterns and visual com-
munication in a group of African monkeys, thus adding a
new perspective to his earlier studies of primate evolution.
He has lectured on this subject in Europe and Japan.

Kilimanjaro: Animals in a Landscape is based on the BBC
television programme 'From Aardvark to Zebra. A Kilimanjaro Sketchbook' in *The Natural World* series.

JONATHAN KINGDON

KILIMANJARO: ANIMALS IN A LANDSCAPE

BRITISH BROADCASTING CORPORATION

Published by the British Broadcasting Corporation
35 Marylebone High Street, London W1M 4AA

This book was designed and produced by
The Oregon Press Limited, Faraday House,
8–10 Charing Cross Road, London WC2H OHG

ISBN 0 563 20230 0

First published 1983

Design: Gail Engert

Phototypeset by Oliver Burridge and Company Limited,
Crawley, Sussex, England
Printed by Acanthus Press Limited,
The Trading Estate, Wellington, Somerset TA21 8ST
Bound by West Country Binders, Weston-super-Mare, Somerset

**Unless otherwise stated all the illustrations
are from the author's collection**

Illustrations:
FRONT OF JACKET **'Clouds, elephants, Kilimanjaro'**
BACK OF JACKET **'Roller theme'**

FRONTISPIECE **'Souvenir of Lolui'. Lolui, a rocky island
in Lake Victoria is inhabited by innumerable birds.
Prehistoric man left his marks in bold paintings on the
boulders and his music can be played on deserted rock gongs.**

INTRODUCTION

My eye was caught by a series of huge black and white striped panels painted on a stable hoarding. I asked 'Why on earth . . .?' and was told about an imaginative naturalist from East Africa who was curious about zebras' responses to a super stimulus of stripes. I was intrigued and then discovered Jonathan Kingdon's magnificent volumes on East African mammals which revealed the work of an artist with acute powers of observation and a zoologist of great originality – what a marvellous subject for a film! I rang him up and asked if he'd ever considered exploring his ideas on television, and he leapt into the making of the programme. I had to choose a mere handful of topics from a lifetime's work, and reveal the richness and complexity of Jonathan Kingdon's way of looking at the natural world. Every day there would be new ideas for the film; he saw elephants wallowing in a swamp which when juxtaposed with a monumental skull, illustrated his own perspectives on the evolution of elephants; a lilac-breasted roller twisting and turning in flight produced a painting with no top and no bottom, so what better than to let it revolve in the film? When he stopped the film during editing, a frozen frame of a snarling cheetah was as exciting to him as the animal in action, and were the stimulus for a series of paintings in which he challenges our vision of facial spots and stripes.

For Jonathan Kingdon there is no clear-cut line between art and science, and with brush and pen he transforms the ordinary into the extraordinary. Both the film and the book reflect his eclectic approach, and they complement each other – the film whets our appetite while the book lets us ponder over some of his ideas and pictures, opening our eyes to new ways of seeing. For without a doubt 'his work doesn't finish here – it's just a beginning.'

CAROLINE WEAVER

Woodland kingfisher (Ardea London, Peter Steyn)

KILIMANJARO: ANIMALS
IN A LANDSCAPE

My involvement with animals goes back to childhood. I grew up in one of the most isolated areas of Africa – up-country Tanganyika, in what is now part of Tanzania.

My playground was peopled by a succession of unexpected arrivals – travellers from the wilderness beyond. Birds and animals that were impossible to ignore.

In Mwanza my nursery window overlooked Lake Victoria. Just outside the mosquito screening a Frangipani bush provided the perch for a woodland kingfisher. In the mornings, sometimes just before dawn, it would wake me with its vibrant shrilling. Half a mile away, in the Provincial Commissioner's garden, another bird replied.

To this day that sound and the flashing of brilliant blue and black wings remain an intense evocation of a distant time and place. Each childhood day, uncharted, unpredictable, began with that firework of colour and noise (colour plates, p. 17).

The kingfisher's muezzin was only the earliest and the loudest of a thousand daily announcements. Each day the senses would be assaulted by messages that may have sounded musical or looked good but were all in languages I couldn't translate. The communities I grew up in were more crowded than any bazaar and I remain an absorbed spectator of the behaviour and diversity of the innumerable characters that jostle one another in the equatorial marketplace. Everywhere there are coded messages, in the shape and colour of a wing, in the cadence of a song and in the rhythm of a flock of birds cutting its arabesque through the sky.

I watch a wandering elephant family weave its eccentric pathways, leaving the marks of its passage in torn grass and trees, pitted footprints and aromatic loaves of dung.

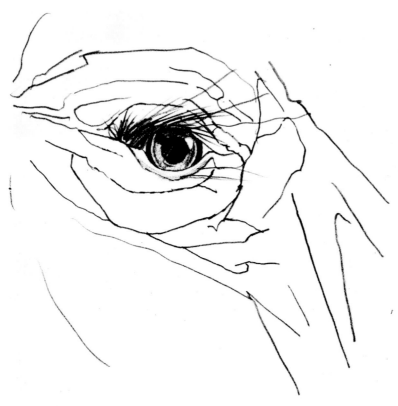

Elephant's eye

These tokens of giant presences are scarcely more obtrusive than the persistence of miniature bees, which savouring the moisture of an eyelid, fog my vision.

Surrounding the immense iris of Kilimanjaro's crater, the mountain's skin folds and creases into valleys where the elephants are tinier than bees.

Blinking from insect to mountain, from bird to elephant, the senses flood the mind with impressions from which a profoundly selective memory builds that private edifice – personal experience.

My sketchbooks record a few such experiences. Impressions of cloud-tumbled skies and mountains are interspersed with studies of insects, jottings, plant drawings and animals in a landscape.

I find the arbitrariness of personal experience a necessary antidote to the impersonal categorisation which science tends to impose upon nature. I am an enthusiastic scientist but I think that 'systems' can acquire an intellectual tyranny which usurps the authority of the senses and can imprison the imagination.

The purpose of this small book, like the television programme from which it derives, is to share a little knowledge and enjoyment of a special place and its inhabitants. Perhaps through shared pleasures we may

The south face of Kilimanjaro

become less foreign to each other and to the world we find ourselves in.

It is in natural history and its offspring, evolutionary biology and field ecology that I have found the search for knowledge most at ease with the evidence of my senses and the wanderings of my imagination. Here are the ideas and methods to sustain pleasure in the minutiae as well as the broad patterns of nature. I have been fortunate in meeting, at appropriate and receptive moments in my childhood, a succession of dedicated and practical naturalists. I have also been fortunate in being taught to observe and draw from life, from about five years of age, by my mother.

I find two aspects of the mind at work whenever I look at nature. One is simply curiosity which asks why are these trees the shape they are or the birds that colour? How did they evolve to be what they are and where they are? What makes one bigger the other smaller? One common the other less so? The other aspect is a desire to celebrate, perhaps even to capture and take home something of an experience. Drawing has been a natural way of both exploring and celebrating natural history. The two impulses can find common ground because there's usually something quite simple and straightforward about an organism's appearance. Its shape, colour, size and setting are both biologically significant and also the very stuff of

9

ABOVE **The spoonbills traverse**

LEFT **Spoonbill (Ardea London, Kevin Carlson)**

imagery. An example: among the commonest and most obvious animal communities are the white herons, ibises and storks which feed in shallow muddy waters and congregate along the shores of lakes and swamps. Sitting on the verandah steps in Mwanza I remember seeing them arrive below the garden and wondered then why that purity of colour and shape?

Later observation led to the conclusion that their whiteness is a vital beacon in their daily search for food. Rapid seasonal cycles of flood and drought means that aquatic fauna tends to be concentrated, ephemeral and unpredictable in its distribution, furthermore it's hidden from view.

I have repeatedly watched birds home in on others feeding successfully: there are rapid build-ups of scores of birds belonging to a dozen or more species. Against the murk of the swamps whiteness signals 'hunt here'.

The pioneers are sacred ibis and spoonbills, whose random search for food is wideranging and almost continuous. The spoonbills walk briskly back and forth, endlessly sweeping their open beaks from side to side through the shallows. The patterns of their movements sometimes remain in the muddy waters like after-images.

A more punctuated trace is left by the sacred ibis as it probes in thicker

ABOVE **Feeding flock of waterbirds: sacred ibises, spoonbills, little egrets and black heron (Ardea London, Gert Behrens)**

RIGHT **Drawing. Birds in a water pan**

mud or more obstructed shallows. Little egrets dart in and out catching disturbed fish and insects; yellow-billed egrets stalk about on the peripheries while the great white egret takes up a strategic position and watches for the ripples of fleeing fish to pass within reach of its long neck.

I have recorded the number of footsteps taken by each species in the course of a minute; spoonbills average sixty or more, sacred ibis forty, little egret thirty-five, yellow-billed egret twenty and great white egret four.

Movements are feverish, steady, jerky, slow or poised and still; kinked necks vary in their spring-loading; silhouettes differ; all signify details of the birds' biology. In the proportions and tempo of the birds are unexpected visual rhythms.

Most of the other waterbirds are more obviously patterned·in colour. The eye is bewitched by the rapid transformations in a plover as he bows, preens or pirouettes from one activity to another.

Like a buxom dancer flaunting her necklaces the double-banded courser sways rhythmically upright, puffing her breast to show off her necklets; her tail fans a sudden flash of black and white; arching wings expose chest-

Double banded courser (Ardea London, Kevin Carlson)

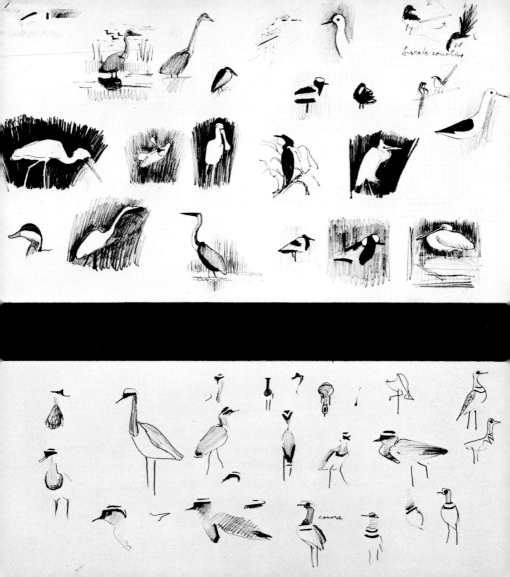

Crocodile

nut patches. Each signal folds away beneath a marbled, dun-coloured mantle and the courser seems to vanish when she crouches.

Lifting the contours of an animal out of its background, identifying its 'significant form' whatever the angle or the lighting is one of the skills to be learnt in drawing. It is also a skill that must have survival value. My own perceptions of the crocodile were definitely sharpened when, at the age of six, I blundered into a small one basking on the lake shore. I know they've never looked like logs to me again.

It is certainly wrong to extrapolate from one's own experience but I would be surprised if Palaeolithic hunters, their eyes searching bushes and plains for particular quarry, did not develop intense search-images of their prey. Could the longed-for animals come to life again in dreams? Would they have materialised in mottled uneven surfaces of cave walls because the visual imagination was still bound up with the search?

I have visited some of the painted rock shelters and caves in Africa and France and seen how contours, details of painted and scratched animals sometimes overlie suggestive bumps and crevices on the stone. It is as

Forest in the valley

though a few marks have given relief to restless eyes and minds by releasing figures trapped in their rock landscape.

Whatever the purposes these drawings served, cave art no less than contemporary drawings are the records of a mind's preoccupations, whether momentary or sustained, playful or compulsive. They are also records of where and how the eye finds pattern.

Over the years my own preoccupations have tilted from the particular portrait or figure in climactic action towards the less obvious overlaid patterns, the slower dynamics of living landscapes.

Trees scrambling over the sides of a valley can be no less exciting than the leap of an antelope or the topography of a human face. In fact they can be rather more challenging. But where to begin? There are so many organisms in a patch of forest, their lives are so deeply interwoven that the extrication of a mere fragment of ecological pattern becomes a triumph for the investigator. From root up to flower every plant has a hidden biography to be untangled. Yet the forest is far from being an aggregate of its parts as a garden is.

However ancient its origins in natural history, the birth of ecology as a science, like the beginnings of landscape painting, was a major intellectual expansion.

Because their origins are so disparate in intention and in time there can be no easy coming together of ecology and the visual arts. Yet I believe that our new and rapidly changing relationship with nature makes it imperative that both the intellect and sentiments of people are touched.

What we observe in the forests and plains around Kilimanjaro differs fundamentally from the farmlands and gardens in which landscape painting originated. In the latter man and animal can be shuffled about at random. In a community animals are the very fabric of a landscape.

There are both historical and structural reasons for landscape painters being more at home in lands that have been shaped by the human will. After all, paintings, like gardens, are given their order by working hands and are contained by the boundaries of a mind.

Cows on their pastures, captives released in parks and food crops in the neighbouring fields are not integral parts of communities, they are transplants. Garden blooms, like cage-birds, have been stolen from their distant niches for the sake of their pretty faces to brighten up the containers we have built for ourselves.

I cannot claim anything very different in bringing back images from the wilderness. Indeed I am fearful that flimsy lines from my own hand, like engravings of quaggas walking the Karroo, may become epitaphs to what once was but will be no more.

Hopefully we don't need to emulate the Cape settlers. New media communicate ecological perceptions that had not been articulated a hundred years ago.

Celebrating the beauty and value of wilderness, interpreting the language and behaviour of animals had to await the advent of colour television before it could achieve widespread appeal.

Predictably television is also becoming the main vehicle for mobilising opinion in favour of conservation. The worst excesses of a voracious international timber trade are more likely to be restrained by outraged television audiences than by the politicians and bureaucrats that have pocketed their votes or fees.

If television can help rescue distant animals in return for delighting its public it can also resuscitate patterns of activity and season that are all but extinct in the cities. The viewer can sit out an hour at the waterhole or

ABOVE **Grants gazelles – territorial ritual** (Ardea London, J. A. Bailey)
OPPOSITE ABOVE **'Drought in Amboseli'**
OPPOSITE BELOW **'Redwing starlings after dusk'**

spend an evening on the rain-soaked plains. Momentarily he may transcend a programme that has been parcelled into parts of seconds and contemplate the irregular rushes and pauses of seasons. He can imagine the urgent imperatives of hormones governing the season of fighting, then cycles of birth and plenty followed by sustained thirst.

Gazelles gather on new flushes of growth on the valley floor. Later the white accents of their elegant buttocks will cut through the dusty heat haze and the drought will be marked out as though it were a territory. The gazelles define the dry season.

High up on Kilimanjaro a short vesper is rung out in the twanging cries and steady flight of red-winged starlings arrowing back to their mountain roosts at nightfall.

There are seasons, hours and minutes in every plant life, every animal cycle. Only a few register their passage in the calendar of memory. Most hover beyond conscious recall.

Leaping impala

I'm sure the Masai, the Egyptians, and many others before and after them, have understood the material metaphors of tempo and season all the better for being without watches.

· A still dark dawn after the night's fast. Falcons off crags in the eastern desert hasten westwards to fill empty crops. They slice through an empty sky, their wings reflecting a sun still below the horizon.

To find the sun's arrival anticipated in the urgent flight of a peregrine is to rediscover an ancient time-mark. Not in a heavy corroded stone or bronze but in a clear fast motion that lifts the spirit as it did in other men four thousand years ago. Horus, divinity of sun and sky, and still killing pigeons for breakfast.

Time and motion. Fifteen thousand years earlier, what were the connotations of galloping horses for the painters of the caves? Who can know? But their descendants have generated more images than any other animal. – In oil-bound ochres, tapestry, bronze, stone and common speech, horses embody speed and power.

While they lost their dignity, first on the battlefields of Crimea and then in the fumes of machines, their poor relations, zebras, bypassed the burdens of human association. Today they suffer from being stars in the National Parks and vermin outside. For Neolithic and modern farmer alike horses

Zebras under Kilimanjaro (Ardea London, John Wightman)

embody gluttony. Paradoxically as real zebras decline their stripes are multiplied by millions – in celluloid and print. Each year the mini-buses halt long enough for twenty thousand camera-happy novices to get their focus right, click, then on to thrill to the impala.

A mother's voice explains 'and God made them all for your delight'. An odd conceit, but who isn't exhilarated when impala take off in a cascade of leaping, jinking bodies? A contemporary sculptor engraves his title 'antelope ballet'. Bushmen artists preferred heavier theatre in the eland. I find myself plodding behind, labouring the question 'why jump like that?' It's a guess, but I interpret what I see as both the laying of air scent-trails and the following of them. Social orientation; prosaic function for a graceful performance set off by a whirlwind or a sudden scare in the thickets.

Tourist, artist, hunter and scientist each superimpose their concepts on structures that are ultimately beyond the mind. I wish long life to the zebra and impala that they may outlast the theories and the deities and I'm confident that pigeon dung will corrode the solemnity of bronze.

Lolui rockscape

There is an uninhabited island in Lake Victoria, Lolui, where cormorants and snake birds air their wings above murals painted on the granite boulders by long lost human colonists. These heraldic birds announce victory for the trypanosome, agent of the sleeping sickness that attacked the islanders seventy years ago. The survivors fled the tsetse flies and their lethal parasites, since when the forest trees and shrubs have come back.

Like a drift of its scent in the wake of a beast gone by, the ancient people of Lolui left the echo of their music. Up in the tumbled kopjies are huge slabs of flaked granite with well worn abrasions. Stone mallets lie below and solicit the visitor to ring a peal on the rock gongs – *In memorian*.

One day, no doubt, people will recolonise Lolui and perhaps the rock gongs will ring again, rediscovered by children from the new villages.

Different people once loved and thought they owned the land they lived upon. Their marks have been overlaid by our own, made with new technologies and governed by changed intentions. Our grandchildren will recolonise our own homes anew. It's a cycle of redundancy that gets faster and faster.

People or proto-people have come and gone over most of these equatorial lands but what of the places that have never been settled? Such as

Mountain bog – Bigo

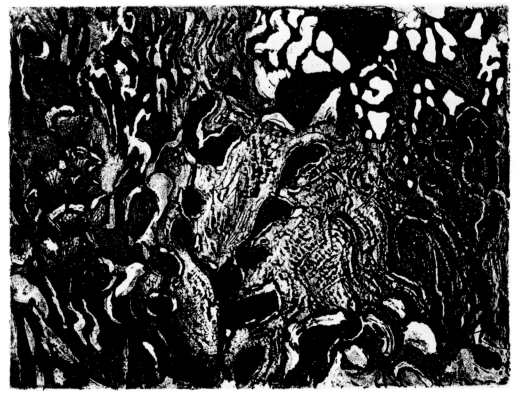

In the moss forest

the upper reaches of the great mountains, Kilimanjaro, Kenya and Ruwenzori.

The last is the wettest and the most extensive; sixty-five miles of peaks, twisting ridges and valleys gouged out by great glaciers of the past. The ancient rocks have been heaved up from below only to tumble down from above, crumbling from crag to boulder, boulder to pebble.

A belt of cloud rings each of the great mountains for most of the day, most of the year. Pretty as a feather boa from a distance, this zone of persistent mists grows sphagnum moss which smothers ground, rocks and tree trunks alike.

Giant heaths dominate this mountain belt; their hard writhing skeletons are eventually embalmed in moss and peat, the living and the dead choke on each other, tangled in a frozen carnival.

To scramble through this gloomy forest is to feel like a beetle seeking its

Giant heaths hung with lichen

way beneath a canopy of unkempt turf. Light from the opaque mists above breaks through in irregular windows.

Where the branches reach open sky the harsh khaki foliage struggles with a dense web of silvery *Usnea* lichen which drapes every exposed surface.

Too much fog and cold, too little sunshine, and that mainly restricted to dawn and dusk, is followed by frosty down-winds from the peaks at night. In a vegetable alliance with mosses and lichens the heaths keep out almost all animal life. The ubiquitous hyraxes and bog rats cannot survive on moss and heath alone. There are a few shrews and the scarce insects sustain just one species of bird. The trilling of an alpine flycatcher comes like the call of a friend through the silent fog.

It is as an animal that one is repelled by a cold hungry darkness that preludes death. As a detached and confident observer there is plenty to

Gall acacia

record and ponder upon. When a shaft of sun comes through, the rippling mounds of moss become exquisite in colour, detail and variety of form. Rocky streambeds rush melt-water down from the snows above. They cut through, exposing the mineral desert that lies beneath the sinuous tangle that has been brushed over moraine and scree. It's absorbing, in its sombre way it's very beautiful but how one misses the fellowship of animals!

Further up are the glaciers. Below the rubble screes floral spires of giant lobelia are landmarks for a vegetation that has pushed right up to its last frontier.

Hidden in a tussock is the highest bivouac. A hollow globular nest woven from the insulating down of everlasting flowers. It is shelter for a malachite sunbird which flits from one lobelia to another searching for minuscule flies and pollinating as it goes.

Plant insect and bird are locked into each other, kindling their own warmths and asserting that life begins and ends here, close to the ice which annihilates life.

In the equatorial lowlands, far below, the points of contact between plants and animals multiply into dense webs and the scale amplifies, culminating in the gigantic masses of mahogany trees and their animal analogues, elephants.

However the rigours imposed by cold on the mountain have their equivalent in fire on the flats. Out on the broad clay-filled valley bottoms what does not drown in the floods dessicates in the drought. What burrows into the clay is cracked open by the sun. What shelters in the red oats will be burnt in the fire.

The major organism able to weather this procession of ordeals is the whistling thorn. Its straggling branches are dotted with black galls and the eccentric stems stand in a halo of long white thorns after the fires have swept all other vegetation away.

On the mountain the sunbird's downy globe holds off the cold. On the scorched plains it is the tough wooden galls of the whistling acacia that withstand fire, sun and the toughest of teeth. Within each hollow sputnik is a hive of ants which have no other home. They defend their capsule with a vigour that augments the tree's other defences, fireproof bark, mouth-proof thorns and crack-proof roots. It is only the policing that would be absent if the balloon-like inflations of its woody stems weren't there. The ants deter all but a few of the plants' many enemies. The tree has no separate existence from the galls or the ants, they have evolved as one.

Spheres are everywhere, each is the container for some vital activity. Cells, eggs, fruits, nests, eyes, brains. Capsules in fibre, tissue, webbing, shell and bone. Carved, spun or built bit by bit. They float free or cluster. They are linked by bridges, arches, struts and buttresses or welded to each other. Space, occupied and contained by boundaries that expand around the nucleus of an organised activity.

In this perspective I see the community that is a whistling thorn as something of a macrocosm for simple cellular organisms and something of a microcosm for much larger communities of animals and plants.

Standing in the smoking devastation that follows a fire they're something to look at and ponder on.

27

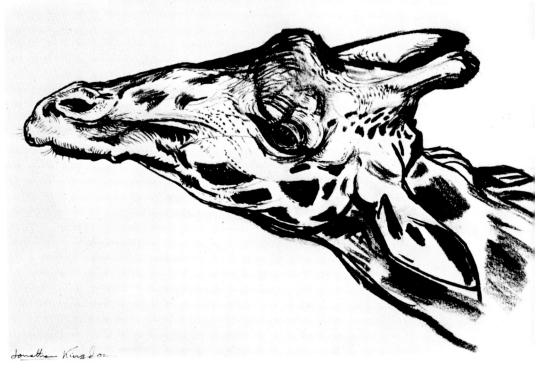

Jonathan Kingdon

Giraffe

The whistling thorn is one of the smaller acacias. Others are among the commonest and most prominent trees of the savannahs.

It is a prominence partly due to their sculptural symmetry. Canopies spread wide, flat-topped in one species, mushroom-shaped in another, tiered and pagoda-like in a third. Tiny feather-like leaflets form a light foliage amidst a dense interlacing of thorny twigs.

Acacias need their thorns. They are legumes and their greenery is a most abundant and richly nutritious type of vegetable fodder.

Acacias need to be fire-resistant. Year after year of their early life the visible tree is decapitated by fire or browsing and year after year the root-stock grows. Eventually the top gets away but even fully grown acacias can suffer a continuous heavy damage and loss of foliage that few other trees would tolerate. They have evolved with the inevitability of these ordeals.

Caterpillars and other insects might eat a lot of the leaves but the

Fever tree and swamp palm

steadiest and most obvious consumers are the elephants and giraffes. The trees are shaped by browsing in more than one way. Green growing surfaces of the tree which can be reached are physically cut back or stripped and this results in contours that reflect where the animals can reach. If the tree is still short it will be a table cropped over its upper surface. If it has got away and expanded above its attacker's head it may be browsed from below or it may be a more vertical structure with its sides indented along the arc of a giraffe's neck-reach, like a baroque vase, but only if the giraffe can walk all round the tree. Without firm standing ground on one side the result is lop-sided pruning.

The tree responds as an individual to continuous cropping from giraffes by a denser regrowth of twiglets, thorns and leaves. Looked at visually acacias are the giraffes' topiary. Looked at ecologically they are the giraffes' market gardening. Regrowth is richer in nutrients than old leaves and the dark eye of the giraffe can be followed as it scans each tree it visits for the latest shoots. It is one long procession by gourmets plucking the acacia equivalent of broccoli tips.

The acacias have responded as species to the certainty that they will be ruthlessly pruned, not only by fire, giraffes and elephants but by many lesser animals as well. Their chemistry, their thorns, their growth patterns, their overall shape, their alliances with ants and other animals, are all evolved features with an ancient history.

That all animals ultimately depend on plants is obvious. Botanical perspectives are more difficult, so perhaps the acacias can help us to visualise how these slow but extraordinarily persistent, active and protean characters are engaged in an endless contest with the elements, with animals and with other plants. Defeat means extinction so it's a curiously ritualised performance in which the rivals are also fighting partners.

We could hardly hope to unravel every nuance of what has gone on but we scarcely try, preferring to blitz our way across tropical lands like illiterate peasant soldiers let loose in a foreign city.

In 1946 the British Government spent thirty-seven million pounds on

Battered acacia trees

clearing acacia bush to grow groundnuts in Tanganyika. Like most schemes, it was the politician's idea, their electors' money; in this case they used demobbed soldiers and military machines to implement it. Where the converted tanks are rusting away today there are still acacias and no groundnuts. They couldn't afford an ecologist.

Twenty years later the United States signed a four million dollar agreement to set up a hundred or so ranches in Uganda. Here too the vegetation and animals were cleared away, forty-four thousand of the larger wild animals were shot. – Rich ranchers and poor peasants have made very bad neighbours.

In the eighties, the schemes, like budgets, are inflated on even more gigantic scales. The cultivator in his field stares in wonder as the machines roll past. I too have stood by watching these immense assaults on the landscape, mutely imagining their busy authors in London and Washington and I think of the unanswered questions in a wilderness left untouched. Each year there are fewer places where it is elephants, not bulldozers, that are knocking the trees about.

Weaver bird colonies – Kilimanjaro (Ardea London, John Wightman)

really blinding
flash on Kili

blinding
glare

from Namalulu gate

They are the only sort of places where we can realistically contemplate some sixty million years of our own origins.

They are places that challenge our ideas of time, of community, habitat, environment. It is the imagination too that is challenged.

My own efforts have centred on attempts to portray the virgin landscapes that I grew up in.

Books have filled over the years with random notes and sketches, people, animals, plants, landscape – a few paintings. Then two invitations led to a deliberate focus on Kilimanjaro.

A visit to Japan prompted me to present a humble homage to Hokusai with an exhibition in Kyoto entitled *Thirty-six views of Kilimanjaro*.

The sketches and miniatures shown there were the first exposure of a private celebration of natural history and Kilimanjaro.

The next step was the making of a BBC film in Amboseli. Once again Kilimanjaro was the *leitmotif*.

The subject was to be 'ways of looking' at animals and their habitats. Passive eavesdropping, with the hard-working cameras pretending to be surrogate eyes, was augmented by an active attempt to make something of it all, constructions in many media.

The colour and movement that television can bring to its subject matter is often subverted by climate, accident and optics. The most imposing and vast landscapes are confidently put away in the can but turn out bleached and pathetic on the box. Vegetation, thrilling in its detail and structure, comes out as a square of mottled green carpet. Our focal themes, Kilimanjaro and landscape were obvious victims of this limitation in photography.

The light moves, the spectrum changes, the observer and the observed all move continuously. The eye revels in colour as it flits from minutiae to mountain. So paintings grew out of the making of the film. They were made in response to the unique dimensions of television – they were made not only in colour but they also moved.

In the film a visual 'overview' was provided by a small hill.

Paintings (pp. 35-56) grew out of sketches (among them pp. 30-40) which were made on the top of that hill. The principal image contained the concept of a 360° panorama where detail and vista broke down into flat coloured shapes like plant or animal patterns (p. 56).

The idea of living animals and landscapes being inseparable was contained in another picture of the aerobatic flight of a roller. Mountain and bird are joined in our experience or memory of the place (p. 55 & cover).

33

Sketches of Kilimanjaro: (top left) **The South face – January**; (top right) **From the Northwest**; (below left) **The South face – October**; (below right) **From the West**

In these paintings Kilimanjaro and its skies were not contemplated calmly, slowly. It was viewed as if through speeded-up film, and that's what it really can be like.

In the morning light creeps in like a wet stain from the East. It reacts with the peak, vapours materialise within seconds, they form, reform, smother, then reveal. As the sun swings through its parabola it cooks up one confection after another; boiling, steaming, frothing or hanging in coloured hazes.

Three miles of mountain throws a mighty shadow. So do clouds as high again. Light spears through, picking out a silver bird, an arc of snow. An

OPPOSITE ABOVE 'The elephant's shadow'

OPPOSITE BELOW 'The passage of Time – Dusk'

OVERLEAF LEFT ABOVE 'Giraffe rosette'

OVERLEAF LEFT BELOW 'Cheetah – Facial calligraphy'

OVERLEAF RIGHT ABOVE AND CENTRE 'Lark landscapes'

OVERLEAF RIGHT BELOW 'Quail plumage'

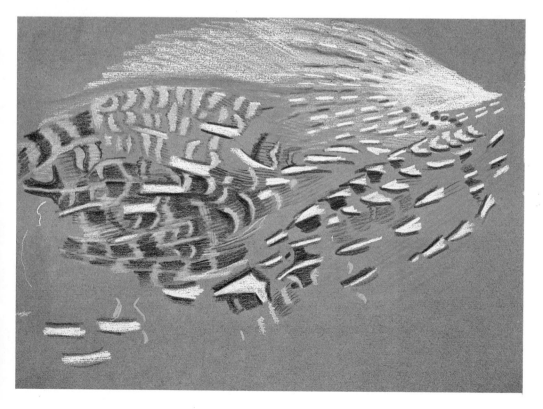

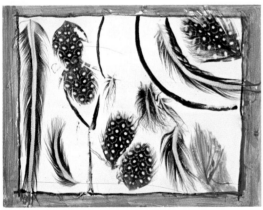

elephant drifts by. Then the layers, the columns, the shafts are all churned. Their hues change fast and suddenly, the night's begun and cooking fires send up their own pale scented cloudlets.

My momentary witness of these vast skies and lands is remembered through little marks and blobs of colour on a fragment of paper. They have less substance than a pebble or a shell, they are more like feathers that are snatched from the wind and kept as a personal *aide-memoire*. This can be taken quite literally. Indeed I have actually used feathers in what could be described as mementos of a passing thought.

The subject (far to the north of the mountain) is vulturine guineafowls. I have picked up three of their feathers, one flask-shaped and spotted, one a streaked spear-head, the third plain azure blue. They tell me something about the nature of the flock around me.

The guineafowls are roughly feather-shaped – each is like a single spotted feather, or should one say the feather is the bird? Suddenly a cock fowl pulls himself up into a tall thin streak and every neck feather becomes an echo of his shape. He twists on tiptoe, pointing to a dropped seed with his scarlet eye, then puffs out his breast, swaying and the blue waistcoat is sensational.

My memory of the guineafowl is made of material supplied to me by the subject itself. The feathers are material, a medium in the literal sense. Looked at biologically it is a material that has been subject to magical transformations devised by evolution. The adult guineafowl is wearing an array of feathers that have derived directly from camouflaged juvenile plumage. This derivation can be traced through the adolescence of a single bird but there are also evolutionary derivations.

The juvenile, like the guineafowl's ancestors, like the quail, (colour plate, p. 37 below) the courser, the lark, the hemipode and a hundred other birds is an exactly scaled representation, a painting in the medium of feathers, of the landscape in which it lives.

Each bird's camouflage owes its similarities to the limited ways in which pigment can be laid down in a feather and to the exact degree to which it shares its neighbour's habitat and body-size. In some of these feathers pigments are in four different tones which occur in identical ratios on any patch of littered ground. The design may coalesce into larger figurations or break down into a filigree to match the uneven tumble of leaves, stems, debris, rocks and sand. The precise effect is to protect these birds from most of their predators.

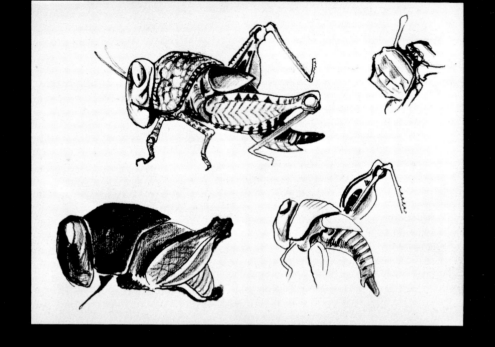

ABOVE **Red capped lark (Ardea London, G. J. Broekhuysen)**

Larks, lost on the open plains, are a sombre but intriguing subject (p. 37). Here's a small place in which the sandy colours are spilt over plumage and rubble alike. Only wings, like the smears and sweeps of a brush, keep their contours – and the eye – always watching.

There are much smaller landscapes further down the scale and they are painted in the medium of insect skin, chitin. Each grasshopper species mimics the exact patterns of light and dark that are found in its own micro-habitat. Ragged bands of colour on the smooth wing covers of one species have roughly the same distance between them as the surrounding pebbles. On the same ground another species is textured like the gravel but in each case the insect is an abstract representation that is tempered by anatomical functions and governed by the physiological limitations of its own body material.

There are other media, snake skin, tortoise scutes, mouse fur, moth wings. Each borrows its ratios and colours from the landscape but each imposes its own order, each medium invents its own particular abstraction. The scales run from the most minuscule insect to the tallest of mammals, the giraffe.

Research has shown that seventy per cent or more of giraffe calves are killed, mainly by lions and leopards, in their first year of life. Like deer fawns the survivors probably owe a lot to markings which are very close in their size and distribution to light and shade in their dry woodland habitat.

Unlike most deer there is no selective pressure to change the coat with age. So the marks enlarge as the animal grows with the end result that an adult pattern is out of scale with its surroundings.

Giraffes and deer are distantly related and a close examination of the pale lines on a Masai giraffe suggests that their ragged edges may be the outlines of pale fawn-like spots which have coalesced to isolate dark 'islands' that we tend to describe as spots or rosettes.

Taking the analogy of a rosette I have plucked two or three blossoms out of the intricate trellice of a giraffe's hide (colour plate, p. 36 above). My picture isolates an abstraction (like formalised stone flowers on a capital or architectural frieze). But I have snatched this image from the actual pattern of a real giraffe.

I have rendered in paint what was itself a re-ordering in fur of landscape elements. For the oxpecker giraffe hides *are* the landscape.

Masai giraffes (Ardea London, John Wightman)

It is commonly assumed that the zebra is camouflaged. A standard zoological textbook which has had its subliminal influence on many zoologists, shows one standing in a clump of black and white bamboo. Other theories are that stripes dazzle the zebra's predators into missing the mark, that the stripes generate droughts to keep the animal cool and that they repel insects.

My own conviction, acquired rather slowly over many years of observation is that the stripes have a social function internal to zebra society. It is they, after all, that have to put up with all that camouflaged, air-conditioned insect repellant. For spotted and blotched zebras to keep on being weeded out (yes, from time to time such aberrations do appear) there has to be some quite fundamental form of natural selection. I suspect that other zebras will exert that selective pressure. Lions, over-heating and tsetse flies are all too erratic and external as influences.

I think the stripes represent a visual bonding system. Imagine growing up within this barricade of black and white. Their dazzle goes with the pleasures of mutual grooming and drinking milk. Need a zebra look further for security and fellowship? They lie where there's dazzle.

Common zebra (Ardea London, Clem Haagner)

Dazzle is shorthand for visual sensations which have been the subject of much experiment. Research has shown that crisp evenly spaced black and white stripes excite at least five primary visual neurones. Therefore animals close enough to get an eyeful of stripes cannot escape being stimulated.

I think that this stimulation may ensure mutually positive responses in any zebra and offset its latent agoraphobia. Aberrant zebras may be selected against by subtle but negative reactions within the herd which serve to reduce the aberrant animal's fitness.

During their travels in search of food and water zebras have to alternate between coming together in very large companies with dispersing into small units. Easy-come, easy-go may appear to be their motto but there is an underlying tension, manifested in frequent snapping and kicking.

As with many other mammals, mutual grooming or nibbling is the main observable way to be friendly but it is a limited, one to one affair. Zebras may have overcome this limitation by associating physical grooming with visual dazzle. I have often watched a zebra nibbling open air, but which, of half a dozen neighbours, was it grooming? Perhaps all of them. If my hypothesis is correct, physical grooming in infancy switches, through a form of conditioned reflex into 'visual grooming'.

There is a climatic requirement for such an optical mechanism to work. Crisp contrasting edges and vivid contrasts need a short sleek coat. Such an elaborate and dangerously conspicuous mechanism could only be justified by a continuous and pervasive social existence. Zebras living at high but variable densities in warm climates satisfy both of these requirements.

Zebras in cold climates need thick coats, which spoils their stripes so they become horses or quaggas. Zebras in deserts have attenuated social life and turn into asses.

Of course stripes are not the only pattern on a zebra. The ears act like railway signals and a rowing action by Grevy's zebra says 'let's go'. Turning round it presents a rump signal which says 'follow me', while a rapidly swishing tail says 'keep off or I'll kick'.

There is general agreement that most of the big cats are well camouflaged. Spattered all over by a spray of inky blobs, the cheetah has blotted out its contours with a pattern that takes its order from the way sunlight splinters through twigs and grasses.

If it is difficult to focus attention on any one part of the spotted body, the same cannot be said for the face. Black and white marks outline the mouth and nose and link up with the eyes. They are quite clearly meant to be

ABOVE **Cheetah head (Ardea London, Clem Haagner)**

seen and the markings serve to emphasise each of the cheetah's very distinctive expressions. Without these accentuations other cheetahs might not only fail to interpret one another's moods but the face itself might get lost in the mosaic of spots.

In the course of filming the cheetah a snarl was caught in full flight. The black squiggles of gums and tear-streak leapt from frame to frame and plate VII on p. 36 is one of several serial images in which the masks have been translated into a sort of Arabic calligraphy which has been made to dance over a field of spots.

The most unlikely animals use visual signals, often startling in their simple geometry.

ABOVE LEFT **Young warthog**

ABOVE RIGHT **Bushpig**

For example, a small movement of the head can convert the horizontal lines of the bushpig's outstretched ears (a posture which is characteristic of a confident, inquisitive pig) into the vertical lines of a submissive or defensive pig with lowered head. Similar whiskers on a young warthog convey quite a different impression. The white mustachios give every appearance of mimicking adult tusks. It's a visual bluff that might work occasionally; in which case it could inhibit attack or injury from pig and predator alike.

Such intimations of visual language are among the incentives to draw but almost any observation can be worth recording in sketch or notebook when it can be put in a framework that has some biological meaning.

48

Pigmy antelope

In the early sixties behaviour, ecology and evolutionary biology had opened up vistas which, for me, were new and exciting. There was also a glaringly obvious need. In spite of people coming from all over the world to see the great animals of East Africa there was not even the simplest form of inventory for the mammals of this region. The enthusiasm of other naturalists encouraged me to embark on a work that was not only a practical inventory and data-bank but also a discussion of the exciting ideas raised by viewing animals in an evolutionary perspective. I subtitled my essay on mammals 'an atlas of evolution in Africa'. Comprehensiveness was not my aim, I knew East Africa was a unique and complex faunal zone. I wanted to advertise the extent and magnificence of the fauna and of the living processes manifest in these animal communities. I wished to put across something of the value of an ancient and little known world which has been integral to human prehistory. With maps I wanted to document destruction of the last great mammals by our civilisation.

Aardvark

Portraying an animal in words and pictures is less an exhibition of skill with language and pencil than a testing of perceptions (one's own and others') of the subject and its setting. Take the aardvark. This is no rag-bag of an animal with a funny name, thrown together from spare parts. It is the largest predator of Africa's most numerous animals – termites. Only a very abundant and reliable food could support such a big, common and widespread animal.

View the aardvark's impact on its surroundings and the countryside can be remapped. The ground is diversified by deep artificial caves, trenches, mounds, scrapes and well-churned soils. You can see it all from the air.

What affects the termites affects the aardvark. Fires, floods, drought and over-grazing, all have their special implications for prey and predator. A termite-aardvark perspective adds another dimension to one's appreciation of a landscape.

In the dry season many African soils are as dense as tarmac; imagine finding and then excavating termites deep within hardened laterite or clay. They must be detected, excavated fast and consumed in bulk, then

Aardvark in a termite mound (Survival Anglia Ltd, Alan Root)

digested efficiently. Each skill requires special equipment that has evolved over some unknown but very great span of time. There is the pulsing blow-drier of a snout with hairy filters in the nostrils. Special olfactory apparatus swells its archaic face, the tongue is long and sticky, the teeth are like calcified cigarette filters and the claws are spade-like.

Today we are familiar with the effect of peculiar clothing, special tools and artificial aids on our perceptions of astronaut, diver or welder. Stripped of their equipment, they stand naked and unimpressive but more recognisably *Homo sapiens*.

I like to contemplate the idea of an essential animal that lurks behind the equipment with which it earns its living.

When the aardvark has been pared away the mammal that is left sheds millions of years of specialisation but is still there, recognisably a primitive mammal. I find this a perspective on evolution that can bring life to fossils, in this case the broken bones of condylarths and hyracoids from the Eocene, some sixty million years ago.

If we were able to be transported to the Eocene, it would be mainly cold-

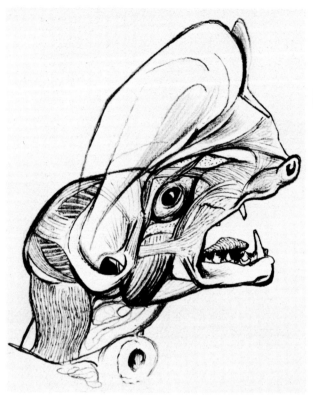

LEFT AND BELOW **Giant mastiff bat**

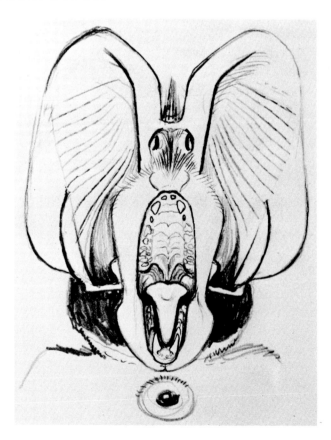

52

Photograph of a giant mastiff bat (Ardea London, P. Morris)

blooded animals that we recognised; like crocodiles. Of the mammals some of the bats would be scarcely different from today but few people would be able to recognise this. Their lives are so far removed from our own, yet they have lived here much longer than we have; they are more diverse and numerous than any other group of mammals and their specialisations have taken them far beyond anything we can find familiar. Even scientific names betray contradictory attitudes. Mystery or fantasy in *Eidolon*, which means ghost but a search for the familiar in 'Molossidae', Greek for bulldog or mastiff. In my dissection drawing of the giant mastiff bat opposite this dog-like element peeps through the bat's more extraordinary adaptations.

WILDTRACK – Jonathan Kingdon Amboseli 16.1.83

The roller not only seems to capture the blue of the sky, and bewitch us with its brilliance, but it also plays with this spectrum. Because it is twisting and twirling through the air there's no top there's no bottom, it is giving us a psychedelic sideshow. One moment presenting some rich lilac pink, the next moment an ultramarine blue and winging it through the air.

It's just not enough to do a picture of the bird, how to try and capture the essence of what that bird has achieved? Through the simple structure of its feathers the roller reflects light just as the sky itself reflects the most wonderful colours, but it has concentrated them and given them very particular shapes. Not fixed shapes, it's not tied to the branch he sits on. It has evolved to be seen in all its variety, movement and dazzle and that I find a very challenging vision. It is the free flashing of one colour against another that offers me the opportunity to put together images which may have something like the concentration and brilliance of colour and also the specific character which I think belongs to the roller and to the place where the roller is experienced.

In this species, which is a fast flier, needing to project its voice loud and far, the lips have flanges like a megaphone. The enormous ears, made more stable by being anchored to the nose, must pick up returning echoes and other voices.

Bats have fascinated me from childhood, everywhere yet nowhere, the sky is alive with their shadowy forms and shrill voices. I find they provoke what Wordsworth called 'obstinate questionings

of sense and outward things,
fallings from us, vanishings;
Blank misgivings of a Creature
moving about in worlds not realised'

I remember being the object of aerial inspection by the largest of African bats. His eyes would reflect my torch-beam as he swooped past, his head craning to see me the better.

Commentary script. Narrator Barry Paine, recording 29.4.83

Kilimanjaro is more than a mountain, it's an experience. Its peak towers three miles above the surrounding plains – this great expanse is broken by an occasional volcanic cone.

Seeking to share with us something of this place, Kingdon finds the cone a vantage point from which to sketch. He plans a picture, but there are a thousand details from which to choose.

The hill-top is a hub, a centre for a panoramic view from which the eye moves out over marshes to the open grasslands and round through woodlands and forests to the great mountain and all that it symbolises.

To commemorate his experience of this rich and diverse landscape, he begins with a series of sketches. The intricacies of nature evoke more than one response in Kingdon. The biologist in him seeks to understand the language of patterns, then out of this view the artist fashions new images.

A panoramic experience, where the leaves of Sansevieria melt in with a stripiness of animals and where, in a dream-like way, the changing colours of the seasons, underlie shapes at once familiar and unfamiliar.

Kingdon has set out to capture the spirit of this living place, its movement, light, colour and the animals that shape it and are in turn shaped by it. All under the distant crescent of Kilimanjaro.

Like *Eidolon*, this magnificent fruit bat has prejudice built into his name: *Hypsignathus monstrosus*, the monster tall-jawed bat. Yet the enormous nose is simply a device to amplify its clanking siren of a voice. Its cheek pouches have spread up until they even hood the eyes. Females are quieter and have faces like labrador dogs but the males are flying loud-speakers, their chests puffed out with trachea like a tuba, throat bulging with resonating sacs and swollen vocal chords.

There are freaks for whom motor-bikes and combine-harvesters are objects of aesthetic contemplation. I sympathise, having found similar satisfaction in a different sort of machine, the warthog. My conversion

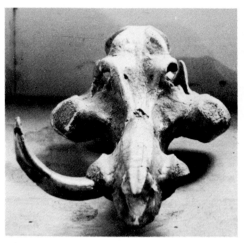

ABOVE LEFT **Hammer bat** (*Hypsignathus monstrosus*)

ABOVE RIGHT **Fossil pig** (*Mesochoerus*)

LEFT **Warthog (Ardea London, Clem Haagner)**

OPPOSITE ABOVE **Drawing for skull sculpture**

OPPOSITE BELOW **Elaboration on a warthog skull**

was clinched by seeing the magnificent skulls of some extinct giant pigs excavated by Richard Leakey's expeditions (they are important for dating hominid fossils).

My own study of the ecology of living pigs, particularly their feeding and fighting, had a bearing on the interpretation of these giant fossils and I was soon embroiled in detailed measurements of the face angles, tusk orientation and tooth structure to help explain why giant hogs and warthogs are the shape they are.

At this time I also needed a relatively complicated model to demonstrate to students what could be done with the lost-wax technique of bronze casting. My current preoccupation with the mechanics of pigs found an immediate (and not very scientific) outlet in a series of bronze pig skulls.

I elaborated on the rhythms of sickle-shaped tusks which swept right back through the skull and I tried to balance each with its own system of thrusts and counter-thrusts. I took considerable liberties with real pig skulls but nonetheless I found myself relying on earlier observation of both the living animals and their bones.

There are numerous discoveries to be made about an animal when one models or sketches from life but preoccupations quickly surface. Invited to join Gregory Maloba, the sculptor, in an aviary of guineafowls, our different predilections soon appeared, his in the choice of a whole bird, my own in a concentration on the head.

The constant movement of living subjects, particularly if they are shy, can make modelling from life a real test of patience. There's nothing to beat it however for getting the 'feel' of an animal in space.

Large lumps of groggy clay can be rather restricting but the odd accretions on a guineafowl's face were my immediate interest so clay was an excellent medium to mould them with. I had not realised, before modelling, how prominent the eyes were, nor had I appreciated the odd angle of the valvular nostrils. In front of muscular cheeks are wattles, as varied in shape and colour as liquorice allsorts but bulging rather horizontally in the bird I was watching. He was also peculiar in the bend to his casque, possibly the product of an injury.

The name, helmeted (or mitred) guineafowl is appropriate. Horny material encases the cranium like a soldier's helmet. This capping may have some semi-physiological or protective function but the larger helmets of males imply that they may have a visual role in reproduction. A threatening male bird pulls himself into a military posture and his horn then resembles nothing so much as an upward pointing beak. It is the same sort of material and shape but bigger and more imposing. The true beak can serve as a weapon but visually is not very impressive, besides there is plenty else to keep it busy with and showing it off wastes energy.

So the horn may resemble the Doge's cap or *corno ducale* in being a ritualised and purely symbolic weapon to reinforce status. Status counts when food and mates are contested in large guineafowl flocks. In the breeding season the birds disperse in pairs to nest discretely in the thickets.

If the guineafowl has reproduced its beak on the back of its head, the hornbills have gone still further. There is one that doubles visual impact by growing an exact replica above the functional one, a pagoda bill or beak-beak.

Another type of elaboration can be illustrated by the yellow-billed horn-bill, which lives in dry woodlands. Long-tailed, gangling birds, pairs mark out their joint territory together with dramatic audio-visual displays from the tops of trees. The males' clucking song is deeper than the female's because he has a puce-coloured resonating sac under his beak. Both sexes flare their spotted wings and bow their heads while the effort of calling shakes the whole body and increases the conspicuousness of the display.

The brilliant yellow beak is slightly translucent, like amber: it's a flag as well as being a tool. The male's chin balloon augments this flagging function and, for other species the sac can be more important visually than the bill. The ground hornbill, for example, has bagpipes which not only send a sonorous booming message but also trick out the black head and beak in brilliant scarlet.

From bag-pipes to violin, sound can be amplified in a soft or hard resonator and some of the most disproportionate and elaborate hornbill beaks seem to serve as musical amplifiers. So in addition to being callipers, pick-axes, plastering trowels and probes, many hornbill beaks are prob-ably sonic resonators and nearly all resemble the yellowbill in having a visual function as well. They are waved in advertisement and fiercely stropped or rattled in threat.

Hornbill weather cock

Tool words help us describe what a thing is. The same point can be made visually with junked tools. A tine broken off a hay-mower just fits the bill, a crushed pipe makes an upper casque, they combine to read as a mechanical hornbill, here mounted on an old door post (with eye-sockets suggested by the dowel hole).

Window handles, scissors, plough shares and old guttering can all combine to make convincing beak elaborations.

Mechanical birds are not just a human perception. A crowned crane reared by friends in a small paddock grew up and looked around for a partner. Ignored by the people who reared him he took to dancing in front of a brass tapped stand-pipe that was his own height. The spiggot was meticulously fed with worms and seeds and had to be unblocked every time his pond needed filling. – The mechanical bride flushed out.

Remembering the tradition of using a bird silhouette, the weathercock, to point the wind from church steeples, I have been prompted to make another hornbill model which explores the 'visual grammar' of a yellow-bill's tree-top display.

64

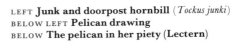
LEFT **Junk and doorpost hornbill** (*Tockus junki*)
BELOW LEFT **Pelican drawing**
BELOW **The pelican in her piety** (Lectern)

There's the arching beak bowing in the same plane above the balloon of its pouch while the flared wings shudder their piebald message.

In a more functional tradition I have also made a chapel lectern to commemorate a Makerere University colleague, William Dickens, who died in the Ruwenzori Mountains. (By a coincidence the Pelican in her Piety is the heraldic crest of his former college.) Why the pelican to symbolise altruism and self-sacrifice? Was it the lance-like beak on snaking neck preening over blood-like iron stains on the belly? Was it the apparent agony of a parent regurgitating to frenzied nestlings which thrust deep

Crowned cranes in flight (Ardea London, P. Blasdale)

into its gullet? The choice of birds in heraldry, ritual and symbolic functions must sometimes have found its origin in some telling detail of natural history that was interpreted or misconstrued by our ancestors. Human history shorn of its animal symbols would indeed be barren and now we need new symbolisms, reminders of our fellowship with them.

The symbolism acquired by birds often has much to do with the ostentation of their displays. One of the most loved of African birds (in spite of its damage to crops) is the crowned crane. Its languid trumpet notes mark dusk and dawn along many an open valley or lake shore. In a flat, monotonous setting the loud calls and colours and the melodramatic movements of cranes are enlarged to become even more obtrusive. It is fitting therefore that the Hima people, who herd their cattle amongst them, should have chosen to imitate the crane in the costumes and movements of their own most elegant dances.

The crane is a dancing animal, made for display and one of its most fascinating aspects is what a choreographer might call its stylistic consistency. I can illustrate this with what can be termed a crane panel (the rather squared-off appearance of the wing, whether open or folded, invites this analogy).

66

'Crane tailoring'

That strong black and white contrast with crisp outlines appears on spanned wings and inflated face and each is a distinct display unit. The golden crest is echoed on the back yet both can fold away beneath serrated grey plumes. These in turn can fluff out and radiate in a pulsing halo (hence the grey stars thrown down on the panel).

Remembering that the Hima found inspiration for their costumes in this plumage I try to organise this panel less as a representation of the crane than as an echo of its tailoring. I try to sum up a few visual characteristics which say 'crane'.

Stylistic consistency can be discerned in many animals. The various guineafowls specialise in charcoal city suits, all types of peacock wear fancy stage costume and entire groups of animals rely on a narrow range of visual effects.

African monkeys are a case in point. Their suits are cut from discrete fabric but there is a geometry in their face patterns which invites questions about the process of speciation. Many faces differ radically from each other but in odd corners of Africa there are populations which suggest both genetic links and hints of the processes whereby the animal differentiated. They stimulated me to make a protracted study which culminated in 1980

L'Hoest's monkey (Phil Agland)

L'Hoest's monkey: ABOVE LEFT **L'Hoest's race;** ABOVE RIGHT **Preuss' race**

in the publication of an essay on the role of visual signals and face patterns. These bold figures are contrived by small alterations of the pigment, relative length of hair and small deflections in the fur's direction of growth. Brightly coloured cowlicks and hair whorls give a sculpture to some of their hair-dos which outpunk the punks.

To separate the visual effects of these patterns from the means by which they were achieved I used various symbolic figures. In one experiment I devised a form of visual Algebra. A small number of rectangles and circles were laid down on a panel and designated roles: let rectangles W and X signify forelimbs, Y and Z the hind limbs. Then a short series of panels were made. In each the proportions and colours of circles and rectangles were altered very slightly but in a definite progression. They got bigger or smaller, darker or lighter, warmer or cooler. The end results were radical transformations of visual effects.

When I used this procedure to compare two related monkeys that live in separate populations in the mountains of Cameroon and Uganda, l'Hoest's monkey's face emerged as a 'target' design. Another species' face resembled traffic lights in blue and yellow. The study showed how pigment is manipulated to make every monkey distinctive.

At the time I was working on this monkey study I was invited to submit designs for a mural in the hallway of one of the large public buildings in Nairobi. I walked back and forth in the concrete cavern wondering if the simple procedures that I was using to explore animal signals might not be

TOP *Cercopithecus l'hoesti*

ABOVE *Cercopithecus l'hoesti* III: **Visual algebra**

70

Cercopithecus l'hoesti II: **Visual algebra**

enlisted both to enhance the architectural spaces and make them distinctive in the way a pattern makes one species distinctive from another that is closely related to it.

People like to picnic beside a waterfall or a pile of boulders because they are landmarks, the focus for a pleasurable activity. The memorability of such places can be reduced to their dominant visual characteristics, a slash of white water, a looming black mass or a blue horizontal. I thought I would try to invest the hall and arcade with a specific sense of place. The chosen material was ceramic tiles in bold flat colours.

So I took the simplest elements in pattern, vertical strips of light and dark, warmer, cooler, set them moving in irregular oscillations and built images out of the interactions when widths changed and edges fractured or slid.

Feathers and butterfly wings were one point of reference, musical analogies another and then there were ancient traditions of decoration. Cultures that weave images out of strips are legion, Jomon whorled goblins, Scythian strapwork stags, Sung interlaced dragons, volutes around Celtic saints, and Islamic floral arabesques to mention but a few. The artists wove

Ceramic mural. I.C.E.A. Building Nairobi, Kenya

their dominant preoccupations into the ordinary fabric of their lives in metals, felt appliqué, painted vellum and tin-glazed terracotta.

Surprise and delight are not easily contained by the methods and practicalities of science. Notwithstanding this the biological world is my central inspiration and a continuous source of renewal; it spills over into daily life. The wilderness has been the larger part of my education and recreation as it has for many generations before.

I think it will always be a world worth celebrating.

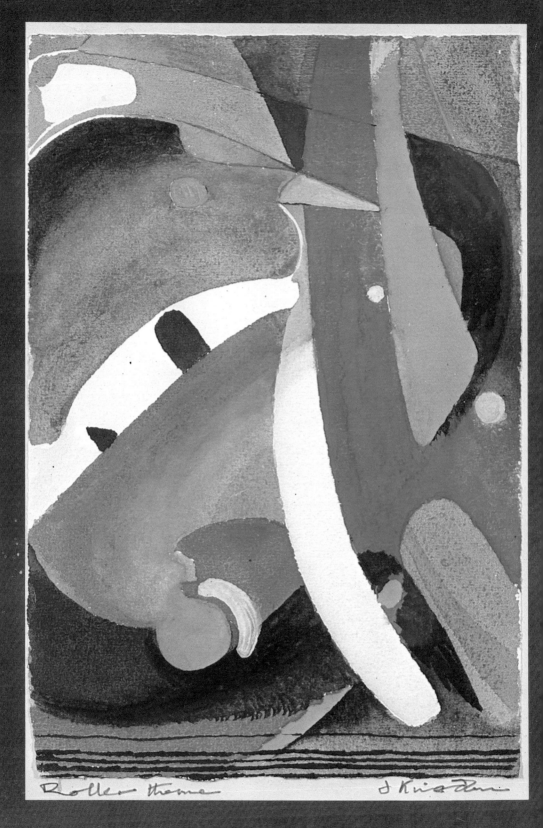

Roller theme   J Kristan

£5.95   ISBN 0 563 20230 0

DATE DUE

DEMCO 38-296